Biochemistry Breakthroughs: Inspiring Stories for Student Scientists

Viktor Petrov

Title: Biochemistry Breakthroughs: Inspiring Stories for Student Scientists

Author's: Viktor Petrov.

This book was printed and published by [Publisher's: Viktor Petrov] in [2023]

ISBN:

TABLE OF CONTENTS

Chapter 1: The Basics of Biochemistry

Introduction to Biochemistry

Biochemistry is a fascinating field that bridges the gap between chemistry and biology. It explores the chemical processes and substances that occur within living organisms, providing a deeper understanding of the fundamental building blocks of life. This subchapter aims to introduce students to the exciting world of biochemistry, inspiring their curiosity and passion for this interdisciplinary science.

Biochemistry plays a crucial role in unraveling the mysteries of life. By studying the structure, function, and interactions of biological molecules, scientists can decipher the intricate mechanisms that drive cellular processes. From the synthesis of DNA to the production of energy in our cells, biochemistry provides insights into the chemical processes that sustain life.

One of the fundamental concepts in biochemistry is the importance of macromolecules. Students will learn about the four major classes of macromolecules – carbohydrates, lipids, proteins, and nucleic acids – and understand how they contribute to the structure and function of living organisms. They will explore the diverse roles of these molecules, such as the energy storage capacity of carbohydrates and the catalytic properties of proteins.

Enzymes, the catalysts of biological reactions, will also be a central topic of discussion. Students will discover how enzymes facilitate chemical reactions by lowering the activation energy, enabling vital processes to occur efficiently. The concept of enzyme specificity, where enzymes interact with specific substrates to form products, will be explored, emphasizing the intricate molecular recognition that underlies biological reactions.

Furthermore, this subchapter will delve into the fascinating world of metabolism. Students will learn about the interconnected pathways that regulate the flow of energy and nutrients in living organisms. They will explore the concepts of anabolism and catabolism, understanding how complex molecules are built and broken down to sustain life.

To enhance their understanding, students will also be introduced to cutting-edge techniques and technologies used in biochemistry research. From X-ray crystallography to mass spectrometry, they will discover the tools that enable scientists to visualize and analyze complex biological molecules, pushing the boundaries of knowledge.

Ultimately, this subchapter aims to ignite students' passion for biochemistry by showcasing its relevance and applications in the real world. By understanding the intricate chemical processes that occur within living organisms, students will be equipped with the knowledge and tools to pursue further research and make groundbreaking contributions to the field of biochemistry.

The Importance of Biochemistry in Scientific Research

Biochemistry, a branch of chemistry that explores the chemical processes within living organisms, plays a crucial role in scientific research. In fact, it is difficult to overstate the importance of biochemistry in advancing our understanding of the natural world and improving human health. This subchapter delves into the significance of biochemistry in scientific research, aiming to inspire student scientists in the field of chemistry.

Biochemistry serves as the foundation for numerous scientific disciplines. By studying the chemical reactions and processes that occur within living organisms, biochemists can unlock the secrets of life itself. This knowledge is then applied to diverse fields, such as medicine, agriculture, and environmental science, to name just a few.

One of the primary reasons biochemistry is instrumental in scientific research is its ability to elucidate the mechanisms behind biological phenomena. From the intricate workings of DNA replication to the complex metabolic pathways that sustain life, biochemistry provides the key to understanding the building blocks of life. By comprehending these fundamental processes, scientists can develop targeted interventions to combat diseases, engineer novel enzymes for industrial applications, and even address global challenges like climate change.

Furthermore, biochemistry plays a critical role in drug discovery and development. Through the study of biological macromolecules, such as proteins and nucleic acids, researchers can identify potential drug targets and design molecules that interact with them. This process, known as rational drug design, has revolutionized the pharmaceutical industry and led to the development of life-saving medications for various diseases.

For students interested in pursuing a career in chemistry, biochemistry offers exciting opportunities for research and innovation. The field is constantly evolving, with new breakthroughs and discoveries occurring regularly. By studying biochemistry, students can contribute to cutting-edge research projects, collaborate with scientists from diverse backgrounds, and make meaningful contributions to scientific knowledge.

In conclusion, the importance of biochemistry in scientific research cannot be overstated. It serves as the bedrock for understanding the complexities of life and provides insights into various scientific disciplines. For students passionate about chemistry, biochemistry offers a fascinating and dynamic field of study, with the potential to make significant contributions to science and society. Embracing biochemistry will open doors to a world of possibilities and inspire the next generation of student scientists.

Fundamentals of Biochemical Processes

Welcome to the subchapter on the fundamentals of biochemical processes! In this chapter, we will dive into the fascinating world of biochemistry and explore the basic principles that govern the intricate processes occurring within living organisms. This knowledge will serve as a solid foundation for your understanding of the field and inspire you to delve deeper into the wonders of biochemistry.

Biochemical processes are at the heart of all life forms, from the smallest microorganisms to complex multicellular organisms like humans. They involve a series of chemical reactions that occur within cells, enabling the organisms to grow, reproduce, and respond to their surroundings. Understanding these processes is crucial for students studying chemistry, as it provides insight into the chemical reactions and transformations that take place within biological systems.

One of the fundamental concepts in biochemistry is metabolism, the set of chemical reactions that transform nutrients into energy and essential molecules. We will explore the intricacies of metabolic pathways, including glycolysis, the citric acid cycle, and oxidative phosphorylation, which are essential for energy production and the synthesis of biomolecules.

Another crucial aspect of biochemical processes is the structure and function of biomolecules. We will delve into the building blocks of life: carbohydrates, lipids, proteins, and nucleic acids. You will learn about the role of enzymes in catalyzing biochemical reactions, the importance of protein folding, and the significance of DNA and RNA in genetic information storage and expression.

To comprehend the dynamics of biochemical processes, it is essential to grasp the concept of equilibrium and kinetics. We will discuss how equilibrium governs reactions within living systems and how enzymes modulate the rate of reactions, ensuring that they occur at the appropriate time and in the right quantities.

Furthermore, we will explore the regulation of biochemical processes, including feedback mechanisms and signaling pathways. Understanding these regulatory mechanisms is vital to comprehend how cells maintain homeostasis and respond to changes in their environment.

By studying the fundamentals of biochemical processes, you will gain a solid understanding of the chemical basis of life. This knowledge will not only lay the groundwork for further exploration in biochemistry but also provide you with a deep appreciation for the intricate processes that drive all living organisms.

So, let's embark on this exciting journey into the world of biochemistry and uncover the breakthroughs that have shaped our understanding of life at the molecular level. Get ready to be inspired as we unravel the mysteries of biochemical processes and their significance in the world of chemistry!

Chapter 2: Pioneers in Biochemistry

Friedrich Wöhler: The Father of Biochemistry

In the world of chemistry, there are several individuals who have made groundbreaking discoveries and revolutionized the field. One such luminary is Friedrich Wöhler, often hailed as the Father of Biochemistry. His remarkable contributions to the study of organic chemistry and the understanding of biochemistry have left an indelible mark on the scientific community.

Born on July 31, 1800, in Eschersheim, Germany, Wöhler displayed a keen interest in science from a young age. His passion for chemistry led him to study under renowned chemist Jöns Jacob Berzelius in Stockholm, where he honed his skills and expanded his knowledge. Wöhler's relentless pursuit of knowledge and his dedication to the field of chemistry eventually led to his groundbreaking discovery.

In 1828, Wöhler achieved a monumental feat that forever changed the course of biochemistry. Until then, it was widely believed that organic compounds could only be synthesized by living organisms. However, Wöhler shattered this notion when he successfully synthesized urea, an organic compound found in urine, from inorganic materials. This groundbreaking experiment, known as the Wöhler synthesis, not only disproved the theory of vitalism but also laid the foundation for the field of organic chemistry.

Wöhler's groundbreaking work in biochemistry opened up new avenues for scientific exploration. It paved the way for the study of the chemical processes that occur within living organisms, providing invaluable insights into the intricate workings of life itself. His discoveries not only revolutionized the field of chemistry but also had far-reaching implications in medicine, agriculture, and pharmacology.

Today, Wöhler's legacy lives on in the countless advancements made in the field of biochemistry. His pioneering work serves as an inspiration to aspiring scientists, particularly those with a passion for chemistry. Wöhler's story exemplifies the power of perseverance and the willingness to challenge established beliefs in the pursuit of scientific truth.

For students interested in pursuing a career in chemistry, Wöhler's story serves as a reminder of the importance of curiosity, dedication, and the courage to question conventional wisdom. His contributions to the field of biochemistry continue to shape the way we understand and explore the intricacies of life.

In conclusion, Friedrich Wöhler, the Father of Biochemistry, stands as a testament to the power of scientific inquiry. His groundbreaking synthesis of urea shattered long-held beliefs and paved the way for the study of biochemistry. Wöhler's story is a source of inspiration for all students passionate about chemistry, encouraging them to push the boundaries of knowledge and make their mark on the scientific world.

Archibald Garrod: Unraveling Inborn Errors of Metabolism

In the realm of biochemistry, there are few names as influential as Archibald Garrod. His groundbreaking work in the early 20th century paved the way for our modern understanding of inborn errors of metabolism. For students passionate about chemistry, Garrod's story is both inspiring and enlightening.

Garrod was born in 1857 in England, and from an early age, he displayed an exceptional aptitude for science. He pursued his education at Cambridge University, where he delved into the world of chemistry. It was during his studies that he became fascinated by the connection between chemical processes and the diseases that afflicted individuals.

Garrod's breakthrough came when he started investigating the concept of metabolic disorders. These disorders, also known as inborn errors of metabolism, are genetic conditions that disrupt the body's ability to convert food into energy or eliminate waste products. Garrod hypothesized that these conditions were the result of specific enzymes being absent or defective.

To prove his theory, Garrod meticulously studied patients with metabolic disorders, such as alkaptonuria and albinism. He analyzed their urine, searching for clues that could shed light on the underlying biochemical abnormalities. Through his research, Garrod discovered that the presence of certain substances in the urine correlated with specific metabolic disorders.

This groundbreaking finding led Garrod to propose the concept of "inborn errors of metabolism," which revolutionized the field of biochemistry. His work laid the foundation for understanding the role of enzymes in metabolic pathways and the genetic basis of these disorders.

Garrod's achievements were not limited to his scientific breakthroughs. He recognized the importance of personalized medicine long before it became a buzzword. Garrod advocated for tailored treatments based on an individual's unique genetic makeup, a concept that is now a cornerstone of modern medical practice.

For students interested in chemistry, studying Garrod's work offers invaluable insights into the intricate relationship between biochemistry and human health. His story serves as a reminder that even the smallest details at the molecular level can have profound effects on our well-being.

In conclusion, Archibald Garrod's unravelling of inborn errors of metabolism is a testament to the power of scientific curiosity and perseverance. His discoveries not only expanded our knowledge of biochemistry but also paved the way for personalized medicine. For aspiring chemists and students of the field, Garrod's story is a testament to the impact that a single individual can have on scientific progress.

Gerty Cori: Breaking Barriers as a Female Biochemist

In the fascinating world of biochemistry, there have been numerous groundbreaking scientists who have shaped the field with their discoveries and contributions. One such remarkable individual is Gerty Cori, a pioneering female biochemist who defied gender barriers and left an indelible mark on the world of chemistry.

Gerty Theresa Radnitz, later known as Gerty Cori, was born on August 15, 1896, in Prague, which was then part of the Austro-Hungarian Empire. From a young age, she displayed an insatiable curiosity for the natural world and a passion for unraveling its mysteries. Despite facing societal prejudices and limited opportunities for women in science, Gerty Cori persevered and became a trailblazer in the field of biochemistry.

Cori's groundbreaking work focused on carbohydrate metabolism, particularly on the study of glycogen. Alongside her husband, Carl Cori, Gerty made groundbreaking discoveries regarding the conversion of glycogen to glucose in the body. Their research led to the elucidation of the Cori cycle, which explained how the body metabolizes glucose and produces energy. This groundbreaking work earned them the Nobel Prize in Physiology or Medicine in 1947, making Gerty the third woman ever to receive this prestigious honor.

As a female biochemist in a predominantly male-dominated field, Gerty Cori faced numerous challenges and prejudices throughout her career. However, she remained undeterred and achieved remarkable success through her determination, intellect, and perseverance. Her accomplishments not only advanced our understanding of biochemistry but also shattered stereotypes and paved the way for future generations of aspiring female scientists.

To students with an interest in chemistry, Gerty Cori's story serves as an inspiration and a reminder that passion and dedication can overcome any obstacle. Her groundbreaking research laid the foundation for further discoveries in carbohydrate metabolism and provided invaluable insights into human physiology. Cori's legacy not only encompasses her scientific achievements but also her unwavering determination to pursue her dreams despite societal barriers.

In conclusion, Gerty Cori's journey as a female biochemist is a testament to the power of breaking barriers and defying societal expectations. Her contributions to the field of biochemistry continue to shape our understanding of human metabolism. By sharing her story, we hope to inspire students interested in chemistry to pursue their passions fearlessly and persistently, knowing that they too can make significant contributions to scientific knowledge and help shape the future of the field.

Chapter 3: Revolutionary Discoveries in Biochemistry

The Structure of DNA: Watson, Crick, and Franklin

The discovery of the structure of DNA is one of the most significant breakthroughs in the field of biochemistry. Understanding the intricacies of this molecule has revolutionized the way we think about genetics, evolution, and human health. In this subchapter, we will delve into the fascinating journey of three scientists – James Watson, Francis Crick, and Rosalind Franklin – who played key roles in unraveling the structure of DNA.

James Watson and Francis Crick are often credited with the discovery of the double helix structure of DNA. In 1953, they proposed a model in which DNA consists of two strands that are twisted around each other in a spiral staircase-like structure. This breakthrough was based on their analysis of existing scientific data and their own X-ray diffraction images, which provided vital clues about the arrangement of atoms in DNA.

However, it is essential to acknowledge the significant contributions of Rosalind Franklin in this discovery. Franklin was an expert in X-ray crystallography and produced high-resolution images of DNA fibers. Her work revealed the helical nature of DNA and provided crucial evidence for the double helix model. Unfortunately, Franklin's contributions were not fully recognized during her lifetime, and her untimely death limited her ability to receive proper acknowledgment.

The discovery of the structure of DNA opened up a new era in biology and paved the way for numerous advancements in genetics and medicine. It enabled scientists to understand how genetic information is stored, replicated, and passed on from one generation to the next. This knowledge has been instrumental in fields such as genetic engineering, personalized medicine, and forensic science.

For students interested in chemistry, studying the structure of DNA offers a fascinating glimpse into the molecular world. It highlights the importance of interdisciplinary research, as scientists from different fields worked together to piece together the puzzle of DNA. It also serves as a reminder of the crucial role that collaboration and knowledge-sharing play in scientific discoveries.

In conclusion, the structure of DNA, as unraveled by Watson, Crick, and Franklin, represents a monumental breakthrough in biochemistry. Its discovery has forever changed our understanding of life and has paved the way for countless scientific advancements. By studying the structure of DNA, students can explore the fascinating world of molecular biology and gain a deeper appreciation for the wonders of chemistry.

Enzymes: Catalysts of Life

Introduction:

In the fascinating world of biochemistry, few concepts are as crucial and awe-inspiring as enzymes. These remarkable biomolecules act as catalysts, driving and regulating essential chemical reactions in living organisms. In this subchapter, we will explore the incredible role enzymes play in sustaining life, shedding light on their structure, function, and the captivating stories behind their discovery.

Unveiling the Enzyme Structure:

Before delving into the realm of enzymes, it is important to understand their structure. Enzymes are typically proteins, composed of long chains of amino acids that fold into intricate three-dimensional shapes. This specific conformation is crucial for their function, allowing enzymes to bind to specific molecules, known as substrates, and facilitate chemical reactions.

The Power of Catalysis:

Enzymes are often referred to as nature's catalysts, as they possess the remarkable ability to accelerate chemical reactions without being consumed in the process. Through a process known as catalysis, enzymes lower the activation energy required for a reaction to occur, thereby making it more efficient and sustainable within a living system.

Enzyme-Substrate Interaction:

The interaction between enzymes and substrates is highly specific, resembling a lock and key mechanism. Enzymes have active sites, which are regions on their structure that precisely fit the shape of the substrate molecules. Once the substrate binds to the enzyme, a series of chemical transformations take place, leading to the formation of the desired product.

Enzymes in Everyday Life:

Enzymes are not only essential for life, but they also have practical applications in various industries. Students studying chemistry will be intrigued to learn about the application of enzymes in the production of biofuels, food processing, and even in the development of pharmaceutical drugs. Understanding how enzymes work paves the way for innovation and sustainable solutions in these fields.

Historical Breakthroughs:

Throughout history, numerous scientists have made groundbreaking discoveries related to enzymes. From the initial observations of Eduard Buchner, who demonstrated that fermentation could occur outside of living cells, to the pioneering work of James Sumner, who successfully isolated and crystallized the first enzyme, these stories of scientific triumph are both inspiring and enlightening.

Conclusion:

Enzymes truly are catalysts of life, driving the chemical reactions necessary for the existence and function of living organisms. In this subchapter, we have explored their structure, function, and the historical breakthroughs that have shaped our understanding of these remarkable biomolecules. As students of chemistry, embracing the world of enzymes opens up a world of possibilities, where innovation and discovery can lead to breakthroughs that benefit both mankind and the environment. So, let us continue to be inspired by the captivating stories of enzymes and embrace the potential they hold for the future of biochemistry.

The Krebs Cycle: Unraveling Cellular Respiration

Welcome, students, to the fascinating world of biochemistry! In this chapter, we will dive into the intricate mechanisms of cellular respiration, focusing on a crucial process called the Krebs Cycle. Prepare to be amazed as we unravel the mysteries of energy production within our cells!

Cellular respiration is the process by which cells convert nutrients into energy. It is a fundamental process that sustains life and allows organisms to thrive. Within the realm of chemistry, understanding the Krebs Cycle is essential for comprehending the intricacies of cellular respiration.

The Krebs Cycle, also known as the citric acid cycle or the tricarboxylic acid cycle, takes place within the mitochondria of our cells. It is a series of chemical reactions that converts acetyl-CoA, a molecule derived from the breakdown of glucose, into energy-rich molecules such as ATP (adenosine triphosphate).

In this subchapter, we will explore the step-by-step journey of molecules through the Krebs Cycle. We will study the key reactions, enzymes involved, and the remarkable transformations that occur. By understanding these processes, you will gain a deeper appreciation for the intricacy of cellular respiration.

Moreover, we will delve into the importance of the Krebs Cycle in generating energy and the overall efficiency of cellular respiration. Discover how the cycle helps produce high-energy molecules and provides the necessary building blocks for other cellular processes.

As student scientists, it is crucial to grasp the significance of the Krebs Cycle in the grand scheme of biochemistry. By unraveling the mysteries of this process, you will be equipped with a solid foundation to explore further biochemical pathways and their implications.

Throughout this chapter, we will present inspiring stories of scientists who made groundbreaking discoveries related to the Krebs Cycle. Their tireless efforts and inquisitive minds have paved the way for our understanding of cellular respiration and its role in maintaining life.

So, get ready to embark on an exhilarating journey into the depths of the Krebs Cycle! By the end of this chapter, you will not only have a firm grasp on the topic but also be inspired to push the boundaries of scientific knowledge. Let's delve into the wonders of biochemistry together!

Chapter 4: Biochemistry in Medicine

Insulin: A Game-Changer in Diabetes Treatment

Diabetes is a chronic condition that affects millions of people worldwide, and its management requires a delicate balance of diet, exercise, and medication. Among the various treatments available, insulin stands out as a true game-changer. In this subchapter, we will explore the discovery, development, and impact of insulin in the treatment of diabetes.

Insulin, a hormone produced by the beta cells of the pancreas, plays a crucial role in regulating blood sugar levels. Its discovery dates back to the early 20th century when scientists Frederick Banting and Charles Best successfully isolated insulin for the first time. This breakthrough laid the foundation for the development of an effective treatment for diabetes.

Insulin works by allowing cells to take in glucose from the bloodstream, thereby reducing blood sugar levels. Prior to the discovery of insulin, diabetes was considered a death sentence. Patients were often restricted to extremely low-carbohydrate diets, which were ineffective in managing the disease's symptoms. Insulin revolutionized diabetes treatment by providing a means to regulate blood sugar levels effectively.

For students interested in chemistry, understanding the structure and function of insulin can offer valuable insights into the complexity of biochemical processes. Insulin is a peptide hormone composed of two polypeptide chains linked together by disulfide bonds. This intricate structure allows insulin to bind to receptors on the surface of cells, initiating a cascade of events that enable the uptake of glucose.

The development of synthetic insulin analogs further enhanced the treatment options for diabetes. These analogs, created through chemical modifications of the insulin molecule, offer improved stability and pharmacokinetics. They can be administered via various routes, including injections and insulin pumps, providing greater convenience and flexibility to patients.

Insulin has transformed the lives of countless individuals with diabetes, allowing them to lead healthier and more fulfilling lives. However, it is important to note that insulin is not a cure for diabetes but rather a crucial component of its management. Ongoing research continues to explore new avenues for diabetes treatment, including the development of advanced insulin delivery systems and the use of stem cell therapies.

In conclusion, insulin has undoubtedly been a game-changer in the treatment of diabetes. Its discovery and development have offered hope and relief to millions of individuals living with this chronic condition. For students interested in chemistry, studying the structure and function of insulin can serve as a fascinating exploration of the intricate world of biochemistry. As the scientific community continues to unravel the mysteries of diabetes, the story of insulin serves as a testament to the power of scientific breakthroughs in transforming lives.

Antiretroviral Therapy: Fighting HIV/AIDS

HIV/AIDS has been one of the most devastating pandemics in modern history, affecting millions of people worldwide. However, thanks to groundbreaking advancements in biochemistry, we now have a powerful weapon in the fight against this deadly virus - Antiretroviral Therapy (ART). In this subchapter, we will explore the incredible impact of ART on the lives of individuals living with HIV/AIDS and how chemistry plays a vital role in this life-saving treatment.

Antiretroviral Therapy is a combination of drugs that target different stages of the HIV life cycle, effectively suppressing the virus and preventing its replication. These drugs work by inhibiting specific enzymes and proteins that are crucial for viral replication, thereby slowing down the progression of the disease. The development of ART required a deep understanding of the viral life cycle and the chemical interactions between the virus and host cells.

Chemistry has played a crucial role in the discovery and development of these life-saving drugs. Scientists have painstakingly studied the structure and function of HIV proteins, allowing them to identify potential drug targets. Through computational modeling and drug design, chemists have developed highly specific antiretroviral drugs that selectively bind to viral proteins, inhibiting their activity. These drugs are often referred to as protease inhibitors or reverse transcriptase inhibitors.

One of the significant challenges in antiretroviral therapy is the development of drug resistance. HIV has a high mutation rate, allowing it to rapidly evolve and develop resistance to drugs. Chemists are constantly working to design new drugs or modify existing ones to overcome drug resistance. By studying the chemical and structural properties of the virus, they can identify potential weak points that can be exploited to develop more effective drugs.

Antiretroviral Therapy has revolutionized the treatment of HIV/AIDS, transforming it from a death sentence to a manageable chronic condition. With proper adherence to the treatment regimen, individuals living with HIV can lead long and healthy lives. However, it is essential to highlight that ART is not a cure for HIV/AIDS. Ongoing research in biochemistry and chemistry aims to develop more potent and less toxic drugs, ultimately striving towards finding a cure.

As students of chemistry, this breakthrough in biochemistry offers a prime example of how our knowledge and skills can be applied to make a real difference in people's lives. By delving into the molecular intricacies of HIV and its interactions with drugs, we can contribute to the development of more effective therapies and potentially contribute to the eradication of this devastating disease.

In conclusion, antiretroviral therapy represents a significant breakthrough in biochemistry, providing hope and improved quality of life for individuals living with HIV/AIDS. Chemistry plays a vital role in the discovery, design, and development of these life-saving drugs. As aspiring scientists, we have the opportunity to contribute to this field and be part of the global effort to combat HIV/AIDS.

Chemotherapy: Targeting Cancer Cells

Chemotherapy is a crucial treatment option in the battle against cancer. It utilizes powerful drugs to target and destroy cancer cells in the body. In this subchapter, we will delve into the science behind chemotherapy and how it specifically targets cancer cells, offering hope to those affected by this devastating disease.

Chemotherapy works by attacking rapidly dividing cells, a characteristic of cancer cells. Cancer cells divide and multiply at an abnormally high rate, causing tumors to grow and spread throughout the body. By targeting these rapidly dividing cells, chemotherapy drugs can halt their growth and ultimately destroy them.

One of the main mechanisms by which chemotherapy drugs target cancer cells is by interfering with their DNA replication. DNA replication is essential for cell division, and chemotherapy drugs disrupt this process. They either prevent the DNA from replicating or cause damage to the DNA strands, preventing the cancer cells from multiplying further.

Chemotherapy drugs can be administered in various ways, including intravenously, orally, or through injections. Once inside the body, these drugs circulate in the bloodstream, reaching cancer cells throughout the body. However, it is important to note that chemotherapy drugs can also affect healthy cells that divide rapidly, such as those in the bone marrow, hair follicles, and the lining of the digestive system. This is why chemotherapy often leads to side effects such as hair loss, nausea, and a weakened immune system.

To improve the effectiveness of chemotherapy, scientists are constantly researching and developing new drugs that target cancer cells more specifically. These targeted therapies aim to minimize harm to healthy cells while maximizing the destruction of cancer cells. For instance, some drugs are designed to specifically target certain proteins or receptors found on the surface of cancer cells, inhibiting their growth and survival.

In recent years, researchers have also made breakthroughs in the field of immunotherapy, which harnesses the body's immune system to fight cancer. This innovative approach uses drugs to stimulate the immune system, enabling it to recognize and attack cancer cells more effectively.

Chemotherapy remains one of the most important weapons in the fight against cancer. By targeting cancer cells and disrupting their growth, these powerful drugs offer hope and a chance for survival to countless individuals around the world. As students of chemistry, it is essential to understand the science behind chemotherapy and its potential for transforming lives.

Chapter 5: Biochemistry in Agriculture and Food Science

Genetic Modification: Enhancing Crop Yield

In recent years, genetic modification has emerged as a revolutionary tool in the field of agriculture, offering a promising solution to address the global challenge of feeding a growing population. By manipulating the genetic makeup of crops, scientists have been able to enhance their yield, making them more resistant to pests, diseases, and environmental stressors. In this subchapter, we will delve into the fascinating world of genetic modification and explore how it is transforming the field of agriculture.

Genetic modification, also known as genetic engineering or biotechnology, involves the alteration of an organism's DNA at the molecular level. Using advanced techniques, scientists can introduce specific genes into the DNA of crops to impart desirable traits. These traits can include increased resistance to pests, improved tolerance to drought or extreme temperatures, and enhanced nutrient content.

One of the most significant applications of genetic modification in crops is the development of insect-resistant varieties. By introducing genes from naturally occurring insecticides found in bacteria, scientists have created crops that can fend off harmful pests without the need for harmful chemical pesticides. This not only reduces the environmental impact of farming but also minimizes health risks associated with pesticide exposure.

Furthermore, genetic modification has also led to the development of crops with improved nutritional content. For instance, researchers have successfully enhanced the vitamin A content in rice, combating vitamin A deficiency in regions where rice is a staple food. Similarly, scientists have engineered crops to have increased iron and zinc content, addressing the prevalent issue of micronutrient deficiencies in developing countries.

While the benefits of genetic modification in crop yield enhancement are undeniable, there are also concerns surrounding its use. Critics argue that genetically modified organisms (GMOs) may have adverse effects on human health and the environment. However, extensive research and stringent regulatory measures ensure that GMOs are thoroughly tested and deemed safe before they are released into the market.

As aspiring student scientists in the field of chemistry, it is essential to appreciate the role of biochemistry in genetic modification. This groundbreaking technology relies heavily on understanding the molecular structure of DNA and the mechanisms by which genes are expressed. By gaining knowledge in biochemistry, you will be equipped with the necessary tools to contribute to the field of genetic modification and make a positive impact on global food security.

In conclusion, genetic modification is a powerful tool that has the potential to revolutionize agriculture and address the challenges of feeding a growing population. By enhancing crop yield through genetic engineering, scientists are paving the way for sustainable farming practices and ensuring food security for generations to come. As students of chemistry, it is crucial to stay informed about the latest advancements in genetic modification and explore the ethical and scientific implications associated with this technology.

Food Additives: Enhancing Shelf Life and Flavor

In the world of food science, one of the key challenges faced by manufacturers is preserving the quality and extending the shelf life of their products. This is where food additives play a crucial role. Food additives are substances added to food products to improve their taste, appearance, texture, and most importantly, to enhance their shelf life. As students of chemistry, understanding the science behind these additives can provide insights into how they work and their impact on our food.

One major benefit of food additives is their ability to prevent spoilage and extend the shelf life of perishable food items. Microorganisms such as bacteria, fungi, and yeasts can cause food to spoil, leading to potential health hazards. Additives like preservatives inhibit the growth of these microorganisms, ensuring that food remains safe for consumption for a longer period.

Furthermore, food additives can enhance flavor and make our meals more enjoyable. Flavors derived from natural sources, such as fruits and vegetables, are often volatile and can degrade over time. Additives like flavor enhancers provide stability to these flavors, ensuring that we experience the same delicious taste every time we consume a particular food product.

It is important to note that not all food additives are created equal. Some additives have raised concerns due to their potential health effects. As responsible consumers, it is essential to stay informed about these additives and make informed choices. Regulatory bodies around the world, such as the Food and Drug Administration (FDA), carefully evaluate the safety of food additives before approving their use. As students of chemistry, we have the power to understand these substances and critically assess their impact on our health.

In conclusion, food additives are a vital component of modern food production. They help preserve the quality and extend the shelf life of our favorite food products while enhancing their flavor. As students of chemistry, we have the opportunity to delve deeper into the science behind these additives and understand their role in food science. By staying informed and making informed choices, we can ensure that we enjoy safe and delicious food for years to come.

Fermentation: From Bread to Beer

Fermentation is a fascinating process that has been used by humans for thousands of years to create some of our favorite foods and beverages. From bread to beer, fermentation plays a crucial role in their production, and understanding the biochemistry behind it can be both intriguing and enlightening for students studying chemistry.

In simplest terms, fermentation is a metabolic process that converts sugars into alcohol, gases, or acids using yeast or bacteria. This process occurs in the absence of oxygen and is responsible for the unique flavors and textures found in various food and drink products.

Let's start with bread. When yeast is added to bread dough, it metabolizes the sugars present and produces carbon dioxide gas as a byproduct. This gas gets trapped in the dough, causing it to rise and giving bread its fluffy texture. The heat from baking then evaporates the alcohol produced during fermentation, leaving behind delicious, freshly baked bread.

Moving on to beer, fermentation is what transforms a simple mixture of water, malt, hops, and yeast into a complex and flavorful beverage. Yeast consumes the sugars present in the malt and converts them into alcohol and carbon dioxide. The alcohol gives beer its intoxicating effects, while the carbon dioxide creates the bubbles we see in a glass of beer. The type of yeast used, along with the specific brewing process, contributes to the wide variety of beer styles and flavors available.

Understanding the biochemistry behind fermentation opens up a world of possibilities for students interested in chemistry. By studying the different enzymes involved in the fermentation process, students can gain insights into how temperature, pH, and other factors affect the final product. They can also explore the role of microorganisms in fermentation and how different strains of yeast or bacteria can produce distinct flavors and aromas.

Moreover, studying fermentation can also connect students to the rich history and cultural significance of food and beverages. Exploring the traditional methods of fermentation used in different cultures can provide insights into the diverse ways humans have harnessed this process to create unique culinary delights.

In conclusion, fermentation is a captivating topic within the realm of biochemistry, especially for students studying chemistry. From the light and airy bread to the complex flavors of beer, understanding the science behind fermentation can help students appreciate the intricate processes that shape our favorite foods and drinks. Whether it's exploring the enzymes involved or delving into the cultural significance, fermentation offers a world of scientific and cultural exploration for aspiring student scientists.

Chapter 6: Biochemistry in Environmental Science

Bioremediation: Cleaning up Polluted Sites

In the realm of chemistry, there is a fascinating field known as bioremediation that has garnered significant attention in recent years. This cutting-edge technique harnesses the power of living organisms to clean up polluted sites and restore the health of our environment. In this subchapter, we will delve into the incredible potential of bioremediation and explore some inspiring stories of its success.

Bioremediation involves the use of microorganisms, such as bacteria, fungi, or plants, to break down or neutralize hazardous substances found in soil, water, or air. These organisms possess unique metabolic abilities that allow them to transform pollutants into harmless byproducts. By employing this natural approach, bioremediation offers a sustainable, cost-effective, and environmentally friendly solution to the problem of pollution.

One remarkable example of bioremediation's effectiveness lies in the cleanup of oil spills. When a devastating oil spill occurs, the harmful effects on ecosystems can be devastating. However, scientists have found that certain bacteria can break down the hydrocarbons present in oil, effectively cleaning up the contaminated area. This discovery has revolutionized the way we approach oil spill cleanups, offering hope for the preservation of marine life and the restoration of affected habitats.

Moreover, bioremediation has proven successful in tackling heavy metal contamination. Toxic heavy metals like lead, mercury, and cadmium pose serious health risks to both humans and the environment. However, through bioremediation techniques, certain plants and microorganisms can accumulate these heavy metals and render them harmless. This approach, known as phytoremediation, has been used to rehabilitate sites contaminated with heavy metals, providing a sustainable solution while avoiding the need for costly and environmentally damaging excavation or incineration.

Bioremediation is not limited to terrestrial environments; it can also be applied to water bodies. For instance, the use of certain aquatic plants and microorganisms has been instrumental in improving water quality by reducing excess nutrients like nitrogen and phosphorus, which contribute to harmful algal blooms and water pollution.

As students interested in the field of chemistry, the study of bioremediation presents an exciting opportunity to explore the intersection of biology, chemistry, and environmental science. Through understanding the principles and applications of bioremediation, you can contribute to the development of innovative solutions for addressing pollution and protecting our planet.

In conclusion, bioremediation stands as a promising field for the cleanup of polluted sites. By harnessing the power of living organisms, we have the potential to restore the health of our environment and mitigate the harmful effects of pollution. As student scientists, you have the opportunity to contribute to this inspiring field and make a lasting impact on the world around us.

Photosynthesis: Nature's Solution to Global Warming

Photosynthesis, the remarkable process that occurs in plants, is not only crucial for their survival but also plays a significant role in combating global warming. Understanding this natural phenomenon is essential for students studying chemistry, as it unravels the intricate mechanisms by which plants convert sunlight into energy while mitigating the effects of climate change.

At its core, photosynthesis is a biochemical process that utilizes solar energy to convert carbon dioxide and water into glucose and oxygen. Through this mechanism, plants act as natural carbon sinks, absorbing carbon dioxide from the atmosphere and releasing oxygen as a byproduct. This process helps to regulate the levels of greenhouse gases, thus mitigating global warming.

In this subchapter, we delve into the fascinating world of photosynthesis, exploring the chemical reactions and molecular components involved. We will explore the role of chlorophyll, the pigment responsible for capturing sunlight, and the electron transport chain, which converts this light energy into chemical energy in the form of adenosine triphosphate (ATP) and nicotinamide adenine dinucleotide phosphate (NADPH). Additionally, we will examine the Calvin cycle, where carbon dioxide is fixed and transformed into glucose.

Furthermore, we will discuss the importance of photosynthesis in maintaining a stable climate. The ever-increasing levels of carbon dioxide in the atmosphere, primarily due to human activities, have led to global warming and climate change. However, the ability of plants to absorb carbon dioxide through photosynthesis acts as a natural solution to this problem. By understanding the underlying chemistry of photosynthesis, students can grasp the significance of this process and explore potential strategies to enhance its efficiency in combating global warming.

Through this subchapter, chemistry students will gain a deeper understanding of photosynthesis and its pivotal role in addressing one of the most pressing challenges of our time – global warming. By appreciating the chemical intricacies of this natural process, students can develop a sense of responsibility as future scientists and explore innovative ways to harness the power of photosynthesis for a sustainable future.

In conclusion, photosynthesis is not merely a biological process but a remarkable chemical reaction that holds the key to mitigating global warming. By studying this natural phenomenon, chemistry students can gain valuable insights into the mechanisms by which plants convert sunlight into energy, while also tackling climate change head-on. As future scientists, it is our responsibility to harness the power of photosynthesis and develop sustainable solutions for a better world.

Nitrogen Cycle: Sustaining Ecosystems

The nitrogen cycle is a fundamental process that plays a crucial role in sustaining ecosystems on our planet. Understanding this cycle is essential for students studying chemistry, as it provides insights into the interconnectedness of living organisms and their environment.

Nitrogen is an essential element for all life forms, as it is a key component of amino acids, proteins, and DNA, which are the building blocks of life. However, nitrogen gas (N_2) in the atmosphere is not directly accessible to most organisms. It needs to be converted into a usable form through a series of biological and chemical reactions.

The nitrogen cycle begins with nitrogen fixation, where certain bacteria and cyanobacteria convert N_2 gas into ammonia (NH_3) or ammonium (NH_4^+), which can be utilized by plants. These nitrogen-fixing bacteria can be found in the soil or in the root nodules of leguminous plants, such as peas and beans.

Plants then take up the ammonia or ammonium ions from the soil and incorporate them into their tissues. This process is known as assimilation. Animals obtain nitrogen by consuming plants or other animals that have incorporated nitrogen into their tissues. Through digestion, the nitrogen is converted back into ammonia, which is then used to synthesize proteins and other nitrogen-containing compounds in their own bodies.

When plants and animals die or produce waste, the organic nitrogen compounds they contain are broken down by decomposers, such as bacteria and fungi. This process is called ammonification, where organic nitrogen is converted into ammonia once again.

The ammonia produced through ammonification can then undergo nitrification, where certain bacteria convert it into nitrite (NO_2^-) and then into nitrate (NO_3^-). Nitrate is the most common source of nitrogen for plants, which can then take it up from the soil and use it to synthesize new proteins and other nitrogen-containing compounds. This completes the nitrogen cycle.

The nitrogen cycle is a delicate balance, and any disruption can have significant consequences for ecosystems. Human activities, such as the excessive use of nitrogen-based fertilizers or the burning of fossil fuels, can lead to an excess of nitrogen in the environment. This can result in eutrophication, where an excessive amount of nutrients in water bodies leads to algal blooms, oxygen depletion, and the death of aquatic organisms.

Understanding the nitrogen cycle is crucial for students studying chemistry, as it provides a foundation for understanding the interconnectedness of living organisms and their environment. By recognizing the importance of this cycle, students can develop strategies to mitigate the negative impacts of human activities and work towards a more sustainable future for our ecosystems.

Chapter 7: Cutting-Edge Technologies in Biochemistry

CRISPR-Cas9: Precision Gene Editing

In the ever-evolving field of biochemistry, one breakthrough has captured the attention of scientists and students alike: CRISPR-Cas9. This revolutionary technology has opened up new possibilities for precision gene editing, offering immense potential in the fields of medicine, agriculture, and beyond. In this subchapter, we will explore the fascinating world of CRISPR-Cas9 and its impact on the field of chemistry.

CRISPR-Cas9 stands for Clustered Regularly Interspaced Short Palindromic Repeats and CRISPR-associated protein 9. It is a technique borrowed from the bacteria's natural immune system, where it acts as a defense mechanism against viral infections. Scientists have harnessed this system to create a precise and efficient gene-editing tool.

At its core, CRISPR-Cas9 works by targeting specific sections of DNA and making precise changes to the genetic code. The Cas9 enzyme acts as a pair of molecular scissors, guided by a small piece of RNA that matches the target DNA sequence. Once the desired gene has been located, Cas9 cuts the DNA strand, allowing researchers to insert, remove, or modify genes with incredible accuracy.

The implications of this technology are vast. In medicine, CRISPR-Cas9 has the potential to cure genetic diseases by editing out faulty genes responsible for various disorders. This breakthrough has sparked hope for patients with conditions like sickle cell anemia, cystic fibrosis, and Huntington's disease, among others. Moreover, CRISPR-Cas9 could revolutionize cancer treatment by targeting and disabling specific genes that promote tumor growth.

Beyond medicine, the applications of CRISPR-Cas9 extend to agriculture and environmental conservation. Researchers are exploring ways to enhance crop resistance to diseases, increase nutritional value, and reduce the need for harmful pesticides. By precisely altering the DNA of plants and animals, scientists can create hardier and more sustainable food sources, potentially addressing global challenges such as food security.

CRISPR-Cas9 has generated significant excitement and debate within the scientific community. While the potential benefits are undeniable, ethical considerations must be carefully addressed. Questions regarding the safety, unintended consequences, and the potential for misuse of this technology have sparked important conversations worldwide.

For students interested in chemistry, CRISPR-Cas9 presents a captivating journey into the world of molecular structure and genetic manipulation. Understanding this breakthrough technology opens the door to a range of exciting career opportunities, from genetic research and pharmaceutical development to environmental conservation and bioengineering.

As we delve into the intricacies of CRISPR-Cas9, it is crucial to remember that this technology is still in its early stages, with much to discover and explore. By inspiring young minds to delve into the world of biochemistry and genetics, we can foster a new generation of scientists who will shape the future of this groundbreaking field.

Mass Spectrometry: Analyzing Biomolecules

Introduction:
In the fascinating world of biochemistry, the study of biomolecules is crucial to unraveling the mysteries of life. Biomolecules such as proteins, nucleic acids, and carbohydrates play essential roles in maintaining the structure and function of living organisms. To understand and analyze these complex molecules, scientists have turned to a powerful tool known as mass spectrometry.

What is Mass Spectrometry?
Mass spectrometry is a technique used to determine the molecular mass and structure of biomolecules. It involves ionizing molecules and measuring the mass-to-charge ratio of the resulting ions. By subjecting the ions to electric and magnetic fields, scientists can separate and analyze them based on their mass and charge.

Analyzing Proteins:
Proteins are the workhorses of life, involved in countless biological processes. Mass spectrometry allows scientists to identify and characterize proteins with remarkable precision. By breaking down proteins into smaller peptide fragments and analyzing their masses, researchers can determine the amino acid sequence of the protein. This knowledge is crucial for understanding protein function and studying diseases caused by protein abnormalities.

Studying Nucleic Acids:
Nucleic acids, such as DNA and RNA, store and transmit genetic information. Mass spectrometry enables scientists to analyze nucleic acids, providing insights into their structure and modifications. By

fragmenting DNA or RNA and measuring the masses of the resulting fragments, researchers can determine the sequence, identify mutations, and even detect epigenetic modifications. This knowledge is essential for studying genetic diseases and developing personalized medicine.

Characterizing   Carbohydrates: Carbohydrates are abundant biomolecules involved in energy storage, cell signaling, and cell adhesion. Mass spectrometry is a valuable tool for characterizing carbohydrates, determining their structure, and identifying modifications. By selectively ionizing carbohydrates and analyzing their fragmentation patterns, scientists can elucidate the arrangement of sugar units and identify glycan structures. This information is crucial for understanding carbohydrate-protein interactions and developing therapeutics.

Applications   in   Drug   Discovery: Mass spectrometry is also instrumental in drug discovery and development. By analyzing the mass and structure of small molecules, scientists can identify potential drug candidates, determine their pharmacokinetics, and study how they interact with target biomolecules. Mass spectrometry plays a vital role in drug metabolism studies and is crucial for ensuring the safety and efficacy of pharmaceuticals.

Conclusion:
Mass spectrometry has revolutionized the field of biochemistry and provided invaluable insights into the structure and function of biomolecules. Its applications in protein analysis, nucleic acid sequencing, carbohydrate characterization, and drug discovery are

vast and continue to expand. As students of chemistry, understanding the principles and applications of mass spectrometry in analyzing biomolecules will open doors to exciting research opportunities and inspire breakthroughs in the field of biochemistry.

Next-Generation Sequencing: Decoding the Genome

In the rapidly evolving field of biochemistry, one breakthrough stands out above the rest: Next-Generation Sequencing (NGS). This groundbreaking technology has revolutionized the way we decode the human genome, providing scientists and researchers with unprecedented insights into the complexities of life at the molecular level. For students interested in chemistry, NGS offers a fascinating glimpse into the future of scientific discovery and opens up new avenues for exploration.

Traditional sequencing methods, such as Sanger sequencing, were time-consuming and expensive. They could only sequence small segments of DNA at a time, limiting our understanding of the genome's intricate workings. However, NGS has changed the game entirely. This high-throughput technique allows for the simultaneous sequencing of billions of DNA fragments, providing a comprehensive view of the entire genome in a fraction of the time and cost.

NGS works by fragmenting the DNA into small pieces, which are then attached to a solid surface and amplified. These amplified fragments are then sequenced using fluorescently labeled nucleotides. The sequencing machine detects the emitted light, allowing for the identification of the nucleotide sequence. The data generated is vast and complex, requiring sophisticated computational tools to analyze and make sense of the information.

The impact of NGS on the field of biochemistry cannot be overstated. It has paved the way for groundbreaking discoveries in personalized medicine, cancer research, and evolutionary biology, among other areas. By decoding the genome, scientists can identify genetic variations that contribute to diseases, allowing for more targeted and effective treatments. NGS has also unraveled the mysteries of evolutionary relationships, shedding light on the origins of species and the interconnectedness of all life on Earth.

For students interested in pursuing a career in chemistry, understanding NGS is crucial. It provides a foundation for advanced research and opens up opportunities to contribute to groundbreaking discoveries. By mastering the principles and techniques of NGS, students can become part of a new generation of scientists driving innovation in biochemistry.

In conclusion, Next-Generation Sequencing has revolutionized the field of biochemistry by enabling the efficient and cost-effective decoding of the genome. For students interested in chemistry, NGS represents an exciting frontier for exploration and scientific advancement. By understanding the principles and applications of NGS, students can contribute to the development of personalized medicine, unlock the secrets of evolution, and make significant contributions to the field of biochemistry.

Chapter 8: Biochemistry in Everyday Life

Metabolism: Fueling the Body

In the vast world of biochemistry, one fundamental process reigns supreme - metabolism. This subchapter explores the remarkable journey of how our bodies convert the food we eat into the energy we need to thrive. Welcome to the fascinating world of metabolism, a concept that lies at the intersection of chemistry and biology.

Metabolism is the sum total of all the chemical reactions that occur within living organisms, enabling them to grow, reproduce, and maintain essential functions. It is a complex network of pathways that transform the molecules from our diet into the energy currency of life - adenosine triphosphate (ATP).

Imagine your body as a high-performance engine, constantly in need of fuel to function optimally. Metabolism is the process that ensures a constant supply of that fuel. Through a series of intricate biochemical reactions, carbohydrates, proteins, and fats are broken down into smaller molecules, releasing energy in the process.

Carbohydrates, such as glucose, serve as the primary fuel source for our bodies. As you consume carbohydrates, they undergo glycolysis, a process in which glucose molecules are broken down into two molecules of pyruvate. This step takes place in the cytoplasm of your cells and generates a small amount of ATP.

But the journey doesn't end there. Pyruvate enters the mitochondria, the powerhouse of the cell, where it undergoes a series of reactions known as the Krebs cycle. This cycle further extracts energy from the breakdown products of carbohydrates, generating more ATP molecules in the process.

Proteins and fats can also be metabolized to produce ATP. Proteins are broken down into amino acids, which can enter various metabolic pathways to generate energy. Similarly, fats are broken down into glycerol and fatty acids, which are then converted into acetyl-CoA, a molecule that enters the Krebs cycle to produce ATP.

Metabolism is an intricate dance of chemical reactions, each step orchestrated by specific enzymes. Without these catalysts, the metabolic processes would occur too slowly to sustain life. Understanding the mechanisms behind these reactions has allowed scientists to develop medicines to combat metabolic disorders such as diabetes, obesity, and inherited metabolic diseases.

As student scientists, exploring the world of metabolism opens up a realm of possibilities. You have the opportunity to uncover new insights into how our bodies function, develop innovative therapies for metabolic disorders, or even engineer more efficient metabolic pathways. By studying the chemistry of metabolism, you embark on a journey that has the potential to shape the future of healthcare and improve the lives of millions.

In conclusion, metabolism is the intricate process that fuels our bodies, converting the food we eat into the energy we need to thrive. It is an extraordinary blend of chemistry and biology, where molecules are transformed into ATP, the energy currency of life. By delving into the world of metabolism, you have the chance to make groundbreaking discoveries and contribute to the field of biochemistry, paving the way for a healthier and more sustainable future.

Vitamins and Minerals: Essential Nutrients

In the vast world of biochemistry, vitamins and minerals play a crucial role as essential nutrients for our bodies. As students delving into the fascinating field of chemistry, understanding the importance of these micronutrients is key to unraveling their impact on our overall health.

Vitamins are organic compounds that our bodies require in small amounts to function properly. They act as coenzymes, facilitating numerous biochemical reactions that are vital for our metabolism. Without these coenzymes, our bodies would struggle to carry out essential processes such as energy production, DNA synthesis, and the formation of new cells. Vitamins are generally classified into two groups: water-soluble and fat-soluble.

Water-soluble vitamins, including vitamin C and the B-complex vitamins, are not stored in the body and must be replenished regularly through our diet. They are easily absorbed by water and are quickly excreted, making it important to consume them daily. These vitamins are commonly found in fruits, vegetables, and whole grains, and play a significant role in maintaining a healthy immune system, promoting skin health, and supporting nervous system function.

On the other hand, fat-soluble vitamins, such as vitamins A, D, E, and K, are stored in our body's fatty tissues and liver. They require dietary fat for absorption and can be stored for longer periods. These vitamins are essential for various bodily functions, including vision, bone health, blood clotting, and antioxidant protection.

Minerals, unlike vitamins, are inorganic elements that are necessary for our bodies to function properly. They are involved in numerous biochemical reactions, ranging from muscle contraction and nerve transmission to maintaining the balance of fluids in our bodies. Some essential minerals include calcium, iron, magnesium, zinc, and potassium, among others. These minerals are required in different amounts, and deficiencies or imbalances can lead to serious health issues.

As student scientists, it is crucial to grasp the importance of vitamins and minerals in our daily lives. By understanding the roles they play in our bodies, we can make informed decisions about our diet and ensure we are providing our bodies with the necessary nutrients. Keeping a balanced and varied diet rich in fruits, vegetables, whole grains, and lean proteins will help us obtain an adequate intake of these essential nutrients.

In conclusion, vitamins and minerals are essential nutrients that our bodies require for optimal functioning. As chemistry students, understanding the role of these micronutrients in biochemistry is pivotal to appreciating the intricate workings of our bodies. By maintaining a well-rounded diet and ensuring we meet our recommended daily intake, we can support our overall health and pave the way for future scientific breakthroughs.

pH and Buffers: Balancing Chemical Reactions

Understanding the concept of pH and buffers is crucial in the field of chemistry, particularly in the realm of biochemistry. In this subchapter, we will delve into the fascinating world of balancing chemical reactions through pH and buffers.

pH, or potential of hydrogen, is a measurement of the acidity or alkalinity of a solution. It is a logarithmic scale ranging from 0 to 14, with 7 being neutral. Anything below 7 is considered acidic, while values above 7 are alkaline or basic. The pH scale plays a vital role in many biological processes, as even slight changes in pH can have profound effects on living organisms.

One of the key factors influencing pH is the presence of buffers. Buffers are substances that help maintain a stable pH by resisting changes when acids or bases are added. They achieve this by either accepting or donating protons to prevent drastic shifts in pH levels. This ability to stabilize pH is vital for biological systems to function optimally.

Buffers are composed of a weak acid and its conjugate base, or a weak base and its conjugate acid. When an acid is added to a buffer solution, the weak base component of the buffer binds to the excess protons, preventing a significant decrease in pH. Conversely, when a base is added, the weak acid component of the buffer combines with the hydroxide ions, preventing a drastic increase in pH. This delicate balance is crucial for maintaining the right conditions for biological reactions to occur.

Understanding pH and buffers is of utmost importance in biochemistry because many biological reactions are pH-dependent. The proper functioning of enzymes, for example, relies heavily on maintaining an optimal pH environment. Enzymes are protein molecules that act as catalysts, speeding up chemical reactions in cells. However, enzymes have specific pH ranges at which they operate most efficiently. Any deviation from this range can lead to enzyme denaturation and a loss of biological function.

In conclusion, the concepts of pH and buffers are fundamental to the study of chemistry, particularly in the field of biochemistry. Understanding how pH affects biological reactions and the role of buffers in maintaining a stable pH environment is crucial for students pursuing careers in chemistry and related fields. By grasping these concepts, you will be equipped with the knowledge to unravel the mysteries of the intricate chemical reactions that occur within living organisms.

Chapter 9: Ethical Considerations in Biochemistry

Human Cloning: A Moral Dilemma

As students of chemistry, we are constantly exploring the fascinating world of biochemistry and its breakthroughs. However, there are certain scientific advancements that raise ethical concerns and force us to ponder the moral implications. One such topic that has ignited debates and controversies within the scientific community and society as a whole is human cloning.

Human cloning is the process of creating genetically identical copies of individuals. This groundbreaking technique has the potential to revolutionize medicine and provide solutions to various health issues. Imagine being able to generate organs for transplantation without the need for donors or developing personalized treatments tailored to an individual's genetic makeup. The possibilities seem endless.

However, the concept of human cloning also raises significant moral dilemmas. One of the primary concerns is the violation of human dignity and autonomy. Cloning could lead to a devaluation of human life, as individuals may be seen as mere products to be replicated rather than unique beings with inherent worth and individuality. Moreover, the idea of creating life solely for the purpose of harvesting organs or conducting experiments poses profound ethical questions.

Another concern is the potential for abuse and the creation of a class divide. Cloning technology could be misused to produce individuals with specific traits or enhance certain characteristics, leading to a society divided based on genetic superiority or inferiority. This could undermine the principles of equality and fairness, creating a world plagued by discrimination and inequality.

Additionally, there are profound psychological and emotional implications associated with human cloning. Clones may experience an identity crisis or struggle with a sense of self, as they navigate the complexities of their existence as copies of another individual. This raises questions about the nature of individuality, uniqueness, and the importance of genetic diversity in shaping our society.

While human cloning holds immense potential for scientific progress, it is crucial for us as students of chemistry to critically examine its moral dimensions. As future scientists, we have a responsibility to consider the ethical implications of our work and ensure that scientific advancements align with our shared values and respect for human life and dignity.

In conclusion, the concept of human cloning presents us with a moral dilemma that demands careful consideration. As students of chemistry, it is essential for us to engage in thoughtful discussions, explore alternative viewpoints, and contribute to the ethical discourse surrounding this groundbreaking technology. By doing so, we can ensure that scientific progress is accompanied by a strong moral compass, ultimately shaping a world where scientific breakthroughs and ethical values coexist harmoniously.

Animal Testing: Weighing the Benefits and Ethics

In the world of scientific research, animal testing has long been a subject of controversy. On one hand, it has been instrumental in numerous breakthroughs and advancements in the field of biochemistry. On the other hand, it raises ethical questions and concerns about the treatment of animals. As aspiring student scientists, it is essential for us to understand the complexities and dilemmas surrounding this issue.

Animal testing has played a crucial role in the development of life-saving medications and treatments. Countless diseases, such as cancer, diabetes, and heart disease, have been studied and understood better through experiments on animals. The insights gained from these studies have paved the way for the development of effective drugs and therapies that have improved and saved countless human lives. Without animal testing, it would be challenging to predict the potential effects of new drugs or understand the mechanisms underlying diseases.

However, the use of animals in research raises ethical concerns. Animals experience pain and suffering, and their use in experiments raises questions about their welfare and rights. The ethical dilemma lies in balancing the potential benefits to human health with the well-being of animals. As student scientists, it is crucial to be aware of these concerns and to explore alternative methods that could minimize or replace animal testing.

Fortunately, advancements in technology have led to the development of alternatives to animal testing. In vitro studies using human cells, computer simulations, and mathematical modeling are some of the alternatives that are gaining popularity in the scientific community. These methods offer the potential to reduce the number of animals used in research and provide more accurate results.

As students of chemistry, we have the opportunity to contribute to this ongoing debate. By advocating for and engaging in research that focuses on finding alternatives to animal testing, we can help reduce the ethical concerns associated with this practice. Additionally, we can explore the ethical considerations and regulations surrounding animal testing to ensure that it is conducted in the most humane and responsible manner possible.

In conclusion, animal testing is a complex issue that requires careful consideration of both its benefits and ethics. While it has been instrumental in numerous biochemistry breakthroughs, it also raises concerns about animal welfare. As student scientists, we must weigh these factors and actively contribute to the search for alternative methods that minimize the use of animals in research. By doing so, we can ensure a more ethical and compassionate approach to scientific advancement.

Gene Patenting: Ownership of Life

In the fascinating world of biochemistry, the concept of gene patenting has sparked intense debates and raised thought-provoking questions about the ownership of life. As students delving into the realm of chemistry, understanding the implications of gene patenting is crucial for comprehending the ethical and legal aspects of this groundbreaking field.

Gene patenting refers to the practice of granting exclusive rights to individuals or companies over specific genes or DNA sequences. This means that the patent holder has the sole authority to use, manipulate, or sell the patented gene for commercial purposes. While gene patenting has undoubtedly contributed to scientific advancements, it has also sparked controversies due to the potential consequences it may have on access to healthcare, scientific research, and the natural world.

One of the primary concerns surrounding gene patenting is the potential restriction it poses on scientific research. When a gene is patented, it becomes the property of the patent holder, who can control its use and dissemination. This could hinder other researchers' ability to study and develop potential therapeutics or cures related to the patented gene. The fear is that scientific progress may be impeded, limiting our capacity to find solutions for life-threatening diseases.

Furthermore, gene patenting raises ethical questions about the ownership of life itself. Some argue that patenting genes is akin to claiming ownership of a part of nature or even life forms. This raises concerns about the commodification of biological materials and the potential exploitation of vulnerable populations who may be denied access to life-saving treatments due to their high costs.

However, proponents of gene patenting argue that it encourages innovation and investment in research. By granting exclusive rights to patent holders, it provides an incentive for companies to invest in research and development, ultimately leading to the discovery of new therapies and diagnostic tools. Additionally, gene patenting can also generate revenue for inventors, allowing them to recover the costs associated with their scientific breakthroughs.

As chemistry students, it is crucial to be aware of the ongoing debates surrounding gene patenting. By understanding the pros and cons of this practice, we can contribute to informed discussions and make ethical decisions in our future scientific endeavors. As the field of biochemistry continues to evolve, it is our responsibility to consider the potential consequences of gene patenting on access to healthcare, scientific progress, and the fundamental principles of life itself.

Chapter 10: Inspiring Careers in Biochemistry

Academic Research: Pushing the Boundaries of Knowledge

Welcome to the fascinating subchapter of "Biochemistry Breakthroughs: Inspiring Stories for Student Scientists" titled "Academic Research: Pushing the Boundaries of Knowledge." As students with a passion for chemistry, you are about to embark on a captivating journey into the world of academic research. Get ready to be inspired by the incredible breakthroughs that have shaped the field of biochemistry and transformed our understanding of life itself.

Academic research plays a pivotal role in pushing the boundaries of knowledge in every scientific discipline, and biochemistry is no exception. It is through rigorous experimentation, meticulous observation, and tireless dedication that scientists have made groundbreaking discoveries that have revolutionized our understanding of the molecular mechanisms that govern life.

This subchapter aims to provide you, as student scientists, with a glimpse into the exciting world of academic research and its impact on the field of biochemistry. You will be introduced to the incredible stories of renowned chemists who have shaped the course of scientific history through their groundbreaking research.

From the discovery of the double helix structure of DNA by James Watson and Francis Crick to the development of CRISPR gene-editing technology by Jennifer Doudna and Emmanuelle Charpentier, these stories will ignite your curiosity and inspire you to push the boundaries of your own scientific potential.

Furthermore, this subchapter will highlight the importance of collaboration, interdisciplinary approaches, and the ethical considerations involved in academic research. You will learn how scientists from diverse backgrounds come together to tackle complex problems and make significant advancements in the field of biochemistry.

By delving into the stories of these remarkable scientists, you will gain a deeper appreciation for the power of academic research and its ability to transform our understanding of the world. You will be encouraged to pursue your own research interests, ask bold questions, and challenge conventional wisdom, all in the pursuit of advancing scientific knowledge.

So, dear students, prepare to be captivated by the stories of scientific discovery, the thrill of experimentation, and the boundless opportunities that await you in the world of academic research. Let these inspiring tales of biochemistry breakthroughs ignite your passion for chemistry and empower you to become the next generation of scientific innovators, pushing the boundaries of knowledge for the betterment of humanity.

Pharmaceutical Industry: Developing Life-Saving Drugs

In the fascinating world of biochemistry, the pharmaceutical industry plays a critical role in developing life-saving drugs that have the potential to change millions of lives. This subchapter will explore the significant contributions made by the pharmaceutical industry, highlighting the intricate process of drug discovery and development. For students with a passion for chemistry, understanding this aspect of biochemistry will provide valuable insights into how scientific breakthroughs translate into tangible solutions.

The pharmaceutical industry is an essential pillar of modern healthcare, constantly striving to discover innovative drugs that target various diseases and medical conditions. Through extensive research and development, pharmaceutical companies combine the principles of chemistry, biology, and medicine to create effective treatments. This multidisciplinary approach requires collaboration among scientists, researchers, and medical professionals, making it an exciting field for aspiring chemists.

The process of drug discovery begins with identifying potential targets, such as specific molecules or proteins involved in disease pathways. This initial stage involves rigorous experimentation and data analysis, where chemists play a crucial role in designing compounds that can interact with the identified targets. Through techniques like computer-aided drug design and combinatorial chemistry, chemists can create a vast library of molecules to screen for potential drug candidates.

Once promising compounds are identified, they undergo extensive testing in preclinical and clinical trials to determine their safety, efficacy, and dosage requirements. This stage involves collaboration between chemists, biologists, and medical professionals to ensure the drug's effectiveness and minimize any potential side effects. Chemistry plays a vital role in optimizing the drug's chemical structure, enhancing its stability and bioavailability.

The pharmaceutical industry also invests significant resources in regulatory compliance and quality control to ensure that the drugs developed meet stringent safety standards. This involves rigorous testing, manufacturing, and distribution processes to guarantee that the drugs are safe, effective, and reliable. Chemists play a pivotal role in developing analytical methods and quality control procedures to monitor drug purity, stability, and consistency.

The impact of the pharmaceutical industry's work is truly awe-inspiring. The drugs developed by these companies have transformed countless lives, providing relief and treatment for various diseases, from cancer to infectious diseases and chronic conditions. Through their dedication and scientific expertise, the pharmaceutical industry continues to push the boundaries of biochemistry, finding new ways to combat diseases and improve human health.

For students interested in chemistry, the pharmaceutical industry presents an exciting career path. By pursuing studies in biochemistry and pharmaceutical sciences, aspiring chemists can contribute to the development of life-saving drugs, making a tangible difference in the lives of countless individuals. The fusion of chemistry and biochemistry in this industry offers a unique opportunity to apply scientific knowledge to solve real-world problems and make a positive impact on society.

Biotechnology Startups: Innovations for the Future

In recent years, the field of biotechnology has witnessed a remarkable surge in innovation, thanks to the efforts of numerous startups. These young and dynamic companies are revolutionizing the world of science and technology, with their groundbreaking discoveries and advancements. This subchapter, titled "Biotechnology Startups: Innovations for the Future," delves into the inspiring stories of these startups, providing students with insights into the immense potential and opportunities in the field of chemistry.

Biotechnology startups are at the forefront of developing novel solutions to complex problems, using the principles of chemistry and biology. By harnessing the power of enzymes, DNA, and other biomolecules, these startups are transforming the way we approach healthcare, agriculture, energy, and the environment.

One such startup, BioTech Solutions, has made remarkable strides in developing personalized medicine. By utilizing chemical and genetic profiling, they can tailor treatments to an individual's unique genetic makeup, leading to more effective and targeted therapies. This breakthrough has the potential to revolutionize the healthcare industry, providing hope for patients suffering from previously untreatable diseases.

Another startup, AgriTech Innovations, is focused on sustainable agriculture. Through the development of biofertilizers and biopesticides, they aim to reduce the reliance on harmful chemical inputs. By harnessing the power of microorganisms, they have created eco-friendly alternatives that not only protect crops but also promote soil health. Such innovations are crucial in ensuring food security and environmental sustainability.

In the energy sector, CleanEnergy Solutions is making significant strides in the development of biofuels. By using advanced chemical processes, they can convert plant-based biomass into renewable fuels, reducing our dependence on fossil fuels and mitigating climate change. Their research has the potential to transform the energy landscape and pave the way for a greener future.

These examples are just a glimpse of the incredible work being done by biotechnology startups. The field of chemistry plays a pivotal role in driving these innovations, as it provides a deep understanding of the chemical reactions and processes that underpin biological systems.

For students interested in chemistry, biotechnology startups offer a wealth of opportunities. These startups are constantly in need of talented scientists and researchers who can contribute to their groundbreaking work. By immersing themselves in this rapidly evolving field, students can gain hands-on experience and actively contribute to shaping the future of science and technology.

In conclusion, biotechnology startups are revolutionizing the field of chemistry and biology, with their innovative solutions to global challenges. Their groundbreaking work in personalized medicine, sustainable agriculture, and renewable energy are inspiring stories that highlight the immense potential of biotechnology. For students fascinated by chemistry, these startups provide a unique opportunity to be part of the exciting future that awaits in the world of biotechnology.

Chapter 11: Tips for Student Scientists in Biochemistry

Effective Study Techniques for Biochemistry

Biochemistry Breakthroughs: Inspiring Stories for Student Scientists

Chapter 3: Effective Study Techniques for Biochemistry

Introduction:

Biochemistry is a fascinating field that explores the chemical processes and molecular interactions within living organisms. As a student of chemistry, delving into the world of biochemistry can be both challenging and rewarding. To excel in this subject, it is essential to adopt effective study techniques that not only help you understand the concepts but also enable you to apply them in practical scenarios. This subchapter will provide you with some valuable tips and techniques to enhance your biochemistry study experience.

1. Organize and Review: Biochemistry involves a wide range of complex concepts and interconnected information. Begin by organizing your notes and study materials in a structured manner. Take the time to review previous topics before moving on to new ones, as this will reinforce your understanding and help you make connections between different concepts.

2. Visualize and Simplify: Biochemical pathways and molecular structures can be overwhelming. Visual aids such as diagrams, flowcharts, and concept maps can help

you visualize and comprehend complex processes. Break down complex information into smaller, more manageable parts to simplify your learning process.

3. Practice and Apply: Biochemistry is best learned through practice and application. Solve practice problems and work on sample questions to reinforce your understanding of key concepts. Additionally, try to apply the knowledge you gain to real-life scenarios or research projects, as this will enhance your understanding and make the subject more practical.

4. Collaborate and Discuss: Engage in group discussions and collaborative learning with your peers. Explaining concepts to others and discussing different perspectives can deepen your understanding and help you retain information more effectively. Take advantage of online forums or study groups to connect with fellow biochemistry enthusiasts.

5. Utilize Resources: There are numerous resources available to aid your biochemistry studies. Make use of textbooks, online resources, research articles, and instructional videos to supplement your learning. Additionally, consult your professors or academic advisors for guidance and additional study materials.

6. Stay Organized and Manage Time: Effective time management is crucial for success in any subject. Create a study schedule that allows for regular review and practice sessions. Break down your study material into manageable chunks and allocate

specific time slots for each topic. This will help you stay organized and prevent procrastination.

Conclusion:

By adopting these effective study techniques, you can enhance your understanding and mastery of biochemistry. Remember that consistency and perseverance are vital for success in this challenging field. With dedication and the right study techniques, you can unlock the fascinating world of biochemistry and make breakthroughs of your own. Happy studying!

Research Opportunities and Internships

For students passionate about chemistry and aspiring to make groundbreaking discoveries in the field of biochemistry, research opportunities and internships can be the first stepping stones towards a successful scientific career. These experiences not only provide invaluable hands-on training but also open doors to new perspectives, collaborations, and networking opportunities. In this subchapter, we will delve into the significance of research opportunities and internships, highlighting their benefits and offering guidance on how to find and make the most of these opportunities.

Research opportunities allow students to actively engage in scientific investigations, working alongside experienced researchers, professors, and mentors. By participating in cutting-edge research projects, students get exposed to state-of-the-art laboratory techniques, equipment, and methodologies, gaining practical skills that can set them apart in their academic and professional journeys. Moreover, research experiences foster critical thinking, problem-solving abilities, and creativity, which are essential attributes for a successful career in chemistry.

Internships, on the other hand, provide students with real-world exposure to the industry. These opportunities can be in pharmaceutical companies, research institutions, government laboratories, or academic settings. Internships offer a glimpse into the multifaceted applications of chemistry in various fields, from drug discovery to environmental sustainability. By working alongside professionals in the industry, students gain insights into the day-to-day operations, challenges, and opportunities in the chemistry field.

To find research opportunities and internships, students can start by reaching out to their professors, academic advisors, or mentors. University career centers often have resources and connections to help students secure internships. Additionally, online platforms and databases dedicated to internships and research opportunities can be valuable tools for finding suitable positions. Networking with professionals in the field, attending conferences, and joining relevant student organizations can also open doors to potential opportunities.

Once students secure a research opportunity or internship, it is crucial to make the most of the experience. This involves approaching the opportunity with enthusiasm, a willingness to learn, and an open mind. Actively engaging in the research project, asking questions, seeking guidance, and being proactive can enhance the learning experience and leave a lasting impression on mentors and colleagues. Additionally, students should strive to build connections, attend seminars, and participate in scientific discussions to expand their knowledge and network.

In conclusion, research opportunities and internships provide valuable experiences for students interested in chemistry and biochemistry. These opportunities foster skills, knowledge, and connections that can shape a successful career in the field. By actively seeking and making the most of these opportunities, students can unlock their potential, contribute to scientific breakthroughs, and pave the way for a fulfilling career in chemistry.

Networking and Building a Career in Biochemistry

In the world of biochemistry, networking and building connections can be the key to unlocking exciting career opportunities. Whether you are a student just starting out or a seasoned professional looking to advance your career, establishing a strong network can help you navigate the field of chemistry and open doors to new and exciting possibilities.

Networking is not just about exchanging business cards at conferences or events; it is about building meaningful relationships with individuals who share your passion for biochemistry. These connections can provide valuable insights, mentorship, and even job opportunities. By actively engaging with professionals in the field, you can stay up-to-date with the latest research, discover new areas of interest, and gain invaluable advice on how to succeed in your career.

One effective way to network is to join professional organizations and attend their meetings, workshops, and conferences. These events provide a platform to connect with experts in the field, learn about cutting-edge discoveries, and showcase your own research. Remember to approach these interactions with genuine curiosity and enthusiasm, as this will help you forge meaningful connections.

Additionally, social media platforms have become powerful tools for networking in the digital age. Online communities, such as LinkedIn and Twitter, allow you to connect with professionals around the world and engage in discussions on topics of mutual interest. By participating in relevant conversations and sharing your own insights, you can establish yourself as a knowledgeable and engaged member of the biochemistry community.

Building a career in biochemistry also requires proactive efforts to seek out mentorship. Mentors can provide guidance, offer career advice, and introduce you to their professional networks. Reach out to professors, industry professionals, or other experienced individuals who inspire you and express your interest in learning from them. A mentor can help you navigate the challenges of the field and provide valuable insights into the specific career path you wish to pursue.

Remember, networking is a two-way street. It is not just about taking, but also about giving. Be willing to share your expertise, offer assistance, and support others in their career journeys. By nurturing a diverse network of professionals, you can create a supportive community where everyone can thrive.

In conclusion, networking and building a career in biochemistry go hand in hand. By actively engaging with professionals, joining organizations, utilizing social media, and seeking mentorship, you can establish a strong network that will enhance your career prospects and open doors to exciting opportunities in the field of chemistry. Embrace networking as a lifelong endeavor and watch as it propels you towards success in your biochemistry journey.

Chapter 12: Conclusion

Recap of Key Biochemistry Breakthroughs

Welcome to the subchapter titled "Recap of Key Biochemistry Breakthroughs" from the book "Biochemistry Breakthroughs: Inspiring Stories for Student Scientists." In this section, we will delve into some of the most significant advancements in the field of biochemistry, aimed at igniting the curiosity and passion of students with an interest in chemistry.

1. Discovery of DNA Structure: One of the most groundbreaking breakthroughs in biochemistry was the elucidation of the structure of DNA by James Watson and Francis Crick. Their work in 1953 paved the way for understanding how genetic information is stored and transmitted, revolutionizing the field of molecular biology.

2. Enzymes and Catalysis: Biochemists have made remarkable progress in unraveling the secrets of enzymes, which are essential for countless biological processes. Understanding how enzymes catalyze reactions has opened doors to developing new drugs, designing renewable energy sources, and optimizing industrial processes.

3. Metabolic Pathways: Biochemistry has played a pivotal role in deciphering the intricate metabolic pathways that power living organisms. From the discovery of glycolysis to the elucidation of the Krebs cycle, these breakthroughs have provided insights into energy production, cellular respiration, and the interplay between different molecules within cells.

4. Protein Structure and Folding: The determination of protein structures using techniques like X-ray crystallography and nuclear magnetic resonance has revolutionized our understanding of how proteins function. Students will find the fascinating stories of researchers who unveiled the structures of important proteins like insulin, hemoglobin, and antibodies truly inspiring.

5. Genetic Engineering and Recombinant DNA: The ability to manipulate and engineer DNA has opened up vast possibilities in biotechnology and medicine. The discovery of restriction enzymes and the development of recombinant DNA technology have paved the way for producing life-saving drugs, genetically modified organisms, and personalized medicine.

6. CRISPR-Cas9: One of the most recent biochemistry breakthroughs, CRISPR-Cas9, has revolutionized genetic engineering with its remarkable precision and ease of use. This gene-editing tool has the potential to cure genetic diseases, enhance crop yields, and eradicate pests.

By revisiting these key biochemistry breakthroughs, we aim to inspire students in their pursuit of knowledge and encourage them to explore the exciting world of chemistry. These stories exemplify the immense impact that biochemistry has on our lives and the endless possibilities it offers for future scientific advancements.

Remember, as students, you have the power to shape the future of biochemistry and make your own breakthroughs. So, let these stories ignite your passion and propel you towards a rewarding journey in the fascinating realm of biochemistry.

The Future of Biochemistry: Exciting Possibilities

In the fascinating world of biochemistry, there are countless breakthroughs that have shaped our understanding of life at a molecular level. As students of chemistry, it is crucial to delve deeper into these discoveries and learn from the remarkable minds that have paved the way for modern biochemistry. This subchapter, titled "References and Further Reading," aims to equip you with a comprehensive list of resources that will inspire and enrich your journey as a student scientist.

1. "Lehninger Principles of Biochemistry" by David L. Nelson and Michael M. Cox: Considered a cornerstone in biochemistry education, this textbook provides an in-depth exploration of the fundamental principles and concepts. With clear explanations and illustrative figures, it offers a solid foundation for understanding the intricacies of biochemistry.

2. "Biochemistry" by Lubert Stryer: Another highly recommended textbook, Stryer's "Biochemistry" offers a comprehensive and accessible approach to the subject. It covers a wide range of topics, including enzyme kinetics, metabolism, and molecular biology, providing students with a well-rounded understanding of biochemistry.

3. "The Gene: An Intimate History" by Siddhartha Mukherjee: While not solely focused on biochemistry, this book explores the captivating story of the gene and its impact on life. Mukherjee's engaging narrative skillfully weaves together science, history, and personal anecdotes,

making it a captivating read for students interested in the molecular basis of life.

4. "Molecular Biology of the Cell" by Bruce Alberts et al.: As biochemistry and cell biology are closely intertwined, this textbook offers a detailed exploration of both fields. It covers essential topics such as DNA replication, transcription, and translation, providing a comprehensive understanding of the molecular processes that underpin life.

5. "The Immortal Life of Henrietta Lacks" by Rebecca Skloot: This gripping non-fiction book tells the story of Henrietta Lacks, a woman whose cells were unknowingly used for scientific research, leading to countless breakthroughs in medicine. It not only highlights the ethical implications of biochemistry but also emphasizes the human aspect behind scientific advancements.

Additionally, students are encouraged to explore scientific journals such as Nature, Science, and the Journal of Biological Chemistry. These publications feature cutting-edge research articles and reviews that offer insights into the latest biochemistry breakthroughs.

Remember, the journey of a student scientist is never complete without continuous learning and exploration. By delving into these resources, you will gain a deeper appreciation for the wonders of biochemistry and be inspired to make your own breakthroughs in the field. Happy reading!